EXCEL MACROS AND VBA

2019

A Complete Step-By-Step Guide

For Beginners

Second Edition

2

without permission or backing by the trademark owner. All trademarks and brands within this book are for clarifying purposes only and are owned by the owners themselves and are not affiliated with this document.

Table of Contents

Page intentionally left blank

CHAPTER 1 – INTRODUCTION

Software and programing have made our lives easy when it comes to professional record keeping, reporting and other tasks. It is necessary for any organization, business, or professional to keep track of accounts, figures and data coming to their business as well. Excel worksheets are one of the basic and popular software that is used by the people professionally. Although it is one of the basic options we have these days, it is still easy to use and come up with so many advance options to try out.

Excel worksheets are a complete package for the beginners as well as professionals who want to do book keeping, account management, data recording, evaluation, generating reports, research analytics and more. If you are getting started with it, the software can be an ultimate guideline for you in the future. It helps you to learn the basics that can lead to the ultimate exploration of advanced features and software as well.

Providing the latest options

Most of the people consider Excel as one of the native and basic software that comes with limited features. On average, these people are not so familiar with all the features and wonders of the software that can help them to get some of the amazing work done in no time. Since its launch, Excel has been a complete package for professionals and with the passage of time, it is coming up with all the latest options. Right now, you can deal with larger data sets, make the ultimate report, use numerous formulas and codes to get the work done. It is all about how much you are going to explore and find out about it in general.

Coming up with advanced features

Excel is not just about covering the options but it offers you the advanced features that help to make a real difference. It is coming up with several extensions and ad-ons that allows you to work smart and efficiently at the same time. The latest technology is all about saving time and increasing productivity with accuracy. The software helps us in meeting up these requirements and makes some of the amazing progress as a whole.

Understanding Excel Macros and VBA

These days there is nothing that is not possible to do with your Excel worksheet. All you need to know is the right command and feature to access the required results. Recently, VBA [Visual Basic Applications] have made it easier to use the software applications and get some quick work done. With the help of latest tools, it is quite comfortable to get the desired results in the end.

In addition to VBA, Excel Macros are another utility in the zone that helps you to get a number of functions done automatically using simple commands and codes. It helps you to have a customized version of

Excel worksheets for you that helps in making better progress overall. Before going further, it is necessary to get a proper understanding of both.

Explaining VBA

The Visual Basic for Applications is a programming language developed by Microsoft and now has become the part of almost all the products of Microsoft Office. The language is more or less like any other language we use to communicate in different regions of the world such as Italian, English, French, German or more. In the world of computer and application, the coding languages help up to make the application function as per our requirements and desires.

VBA comes with a set of commands and alphabets with combinations just like any other language that helps to make a manuscript. These manuscripts or codes are actually the commands to software that helps us to get the ultimate results.

Understand Macros

Most of the people relate VBA and Macros as one; these are similar but not the same. There is a fine difference between both that needs to be considered in the first place. Macros are the codes and commands that are designed using the VBA language. These Macros in Excel helps to perform multiple tasks and increase productivity with a significant margin.

To work efficiently with your VBA features in Excel, it is necessary that you should know the basic macro creation and functions too. For the beginners, it is important to learn the creation of the macros properly so they can enhance productivity as a whole.

To get started with the VBA it is necessary for you to learn programing in the first place. It seems to be difficult for some of the people but pay off well. On the other hand, there is good news that programing in Excel is easier than expected. Initially, it feels like to be a hard option to learn the programing but when you practice in contrast, it looks easier.

How Macros are different from VBA?

As it was mentioned earlier, VBA and Macros are not the same things. These can be relatable but we cannot interchange both and use then instead of any one of these. It is important for beginners to understand the difference and its importance so they will get the right results as well.

VBA is a programing language that is used in several Microsoft programs in Microsoft Office. It is used to construct and formulate a program that helps in getting a final product eventually.

On the other hand, Macro is a sequence of instructions that are used to get the results using the VBA language. Professionals do not appreciate confusing macro with a programming language. A pattern helps to achieve the desired results, so it cannot be identified as a language in any case.

In simple words, we use the Visual Basic for Applications as a programming language to create macros. For the Excel Macros, there is a complete set of instructions comes in a sequence. The sequence should be followed in order to get the desired or promised results in the end. We can say that both are well connected but cannot interchange them in places.

In fact, VBA can do nothing without macros and to develop macros, you need the VBA language. Both of these have an inter-dependency on each other and the programmers need to learn both. It starts with the understanding of VBA in the first place and that leads to the learning macros and knowing the pattern. Once a person can understand language and sequence of macros then its quite easy to work with these tools.

Learning the VBA Language

Computer languages are slightly different from normal communication language. They have some of the specific limitations and restrictions in terms of use and presentation. The programing language cannot go advance to the extent that will replace the ears, eyes, mouths, and figures as well. Learning any of the language involves the specific parameters that are considered to ensure the ultimate results.

Whenever you are learning a new communication language, then familiarity of the basics in terms of grammar or composition makes it an easy deal for you. If you do not have any previous experience with the language, then it is quite challenging to learn and manage a new language. The same thing happens in the case of VBA and other programing language. If you have a little knowledge about the programing and coding, then it is quite easy and nice to learn the language.

In terms of structure, all the programming languages are quite similar. They are not identical but the format seems to be familiar that gives you the liberty of understanding the similarities. On the other hand, this structure is not similar to the human communication language, so you need to learn a few basic things in advance. It will help you to make real progress in learning Excel VBA macros.

Uses and advantages

VBA Excel Macros seems to be an extension to the Excel that helps you to work efficiently with the worksheets. It makes things quick, easier and accurate as a whole. Commonly for the complex jobs with Excel, you need to apply the formulas and commands multiple times, but using the macros, you can simply get the job done with one single command. It comprises of a series of advantages that makes your life and job easier as a whole.

Making Excel easier

Excel is known as one of the easy and quick tools when it comes to record keeping and data assembling. No matter if you are using it for record or research purpose, you can make things real quick and easy. VBA and macros have made it easier than before and give you more options to make some quick entries. You just need to get the right understanding of writing macros and you will get the best results.

Getting advanced solutions

Macros and VBA cover all advanced and up to mark solutions that you may not find in random coding. With the simple codes and formulas in Excel, you can make one step progress, such as you can work on a single sheet, column or row. But with the help of macros, you are able to work with the whole worksheet, columns, rows and more. Eventually, it saves your time and energy to make things work.

Easy interface

Apparently, for the non-programmers or newbies, it is hard to understand the programing language. They do not consider that it can be easy and will be such a handful for them. In comparison to other software or coding options, macros are quite easy and reliable. Once you have the understanding about VBA, you can simply have the best hands on skills of macros writing.

Compatible language and coding

The coding and VBA language is widely compatible with all other Microsoft products, including Excel. It makes it effective not only for the Excel sheets but for other programs as well. You can use the language to have further advancement in many different programs and software functions as well. Learning one language will help you to have command over multiple actions and possible software skills in the future.

No extensions required

When you are using the VBA Excel macros, you have direct access to the codes and command. For the application of these codes, you do not need to have any app plugin or extension at all. Most of the times, extensions or plugins cause you big trouble when it is about having some advanced options with a software. A bug in the extension can cause a whole host of application issues. In the case of VBA macros, things are quite smooth and sorted from the initial stage.

Quick and clean

When you are looking for a solution to your problems in no time, then VBA Excel Macros is the ultimate solution you have. The procedure to learn the VBA and write the macros is clean and straightforward. You do not need to have any add on to the basic software. It does not complicate your overall interface of the application so you will end up having a clean and quick solution. Moreover, the result rate is better than before, that reduces the chances of errors.

What you need to learn before the start?

Before getting started with the VBA and macros,

there are some basic tasks you need to do. Remember, wherever you are starting up something new, you need to have some preparations with a basic floor plan. It helps you to get some of the important results by investing your time and efforts in the right zone. If you are not taking the first step first and jump to the second or last one, then you will have to face issues with the overall setup. In the case of VBA and macros, you need to consider the following essential points:

Learn programing

> To get comfortable with the VBA and understand macros, you need to be familiar with programing. Many of the professionals are not very comfortable with programing who use Excel worksheets. The use of worksheets is not limited to the programmers but these are widely used by the commerce and business professionals. So, it is crucial to learn the basic programing that will help you to learn more about the coding and make use

of these codes and files easily.

Know the language

As we have discussed earlier, VBA is the language used in macros to make things work in the end. So, it is necessary to get started with the language that is providing you all the basic framework. When you know the language, you will be able to make a real difference in further progress. All you need is to understand the best of basics and practice it well. The knowledge of the language will help you to identify the use of its codes so you can write macros efficiently. It will lead you to the proper implementation of the macros to give multiple commands in Excel sheets.

Clarify concepts

Before getting, entitle as a pro in using the macros and VBA, you need to have the concept clarity. It is important that you are good enough to understand the basics and make some real differences as well. Make sure you are going to clear out all the possible problems that can bug you during work although there are trouble shoot guides available, but you need to know the basic issues, errors, their causes, and immediate solutions. It helps you to save time and energy. You can make some real quick moves to sort out the problems and it eventually helps you to enhance productivity.

Be friends with codes

If you are not comfortable with programing and codes, then things can be hard for you. When dealing with VBA and macros, you should have the knowledge about the coding and you need to improve your ability to understand the coding. It helps you to get the real job done. Although you can find some specific codes online that, you can keep in your records so you do not have to write them repeatedly but even this requires knowledge about codes. When you know the procedure to write a code and its important components only, then you will be able to get the right codes from the resources.

How to train yourself?

To get started with the Excel Macros VBA, it is important for you to prepare yourself. For the professionals who need to work on a frequent basis with good speed and analysis needs to get some real skills. In this regard, you can select your won training criteria that will help you to take a step forward for the ultimate grip and skillset related to macro coding in VBA. Here are some tips for you while getting training for the VBA and preparing yourself to be good enough.

Be patient

Whenever we are starting up to learn something new, there are a number of things that comes in our way. We may not be able to understand it initially, practice will be difficult, and there can be errors and more. Although VBA is comparatively an easy language and with a little of programing knowledge, you can learn well out of it. However, initially, you might face failures or errors with the codes or macros. Therefore, you need to be patient in the procedure so you will not lose the potential of making the best out of your learning.

Start from the easy codes

It is a bad idea of jumping to the complex codes or macros in the beginning. It will do no help but exhaust your brain processing power and leaves you drained. You need to start with the basics of the language and get the hands on experience on easy codes in the first place. These codes will help you to understand the basics of macro writing and its essential codes as well. Later on, you will be able to make some of the complex and customized macros using these basics and achieve the best of

Excel macros functions.

List out the basic examples

During the training procedure, you need to
collect resources and need to have some good
examples. These can be the basic examples so
you can try them and have the ultimate hands
on experience. Eventually, you can end up
having the ultimate expertise with these
examples that will lead you to the professional
command on the macros and more. To keep
these examples streamed, it is an idea to list
them well so you will have clear access to the
examples whenevernyou want to get a
reference.

Save the codes

You do not have to memorize the codes
yourself; it is not possible all the time. Instead
of remembering, they make sure to save them
with you. Keeping a file of codes on your
server will help you in making quick macros
and get the job done in no time. It helps you

28

to have better and advanced skills in macro writing and creating the best codes to get your work done.

Make and crack your codes

It is not necessary that you stick to some of the scripted codes you have got in the beginning. Training yourself with the VBA and Excel macros means you are supposed to make some of the advancements in it. So, you need to expand the expertise and make some of the important moves with establishing new skills in the program. Furthermore, once you have the idea about how to write macros and what command functions what, you can easily move forward with your own customized codes and macros.

Keep it simple!

Remember, whenever you are doing something new, then you need to start with some simple steps initially. If you try to make it harder or complicated in the beginning, things will go out of control. It is necessary for you to keep things in control and easy for you, so it is ideal to start with the basics and easy small steps and move gradually towards the complex and larger actions. It will not only helps you to have the ultimate grip on codes and language but you can let other people learn it from you.

CHAPTER 2 – GETTING STARTED

The use of VBA and macrons for the users of Excel spreadsheet make work easy and quick. A powerful language and tool can be extended and helps to accomplish great tasks for its users. People are interested to learn about these because it can be customized as per the use of the application. Through macros, you can perform a task, alter file, data, information, and format or arrange spreadsheets and get useful documentations with no time. With the personalized macros, you do not need to search for the codes through this, you can make your own bar by creating and highlighting macros or codes that can make your task easier. Macros can help with the editing, text styles, to add column, row, and highlight, add or delete columns, rows, and much more.

Quick start by recording macro

A person with a little knowledge and exploration can not only create but also make customize or personalized macrons list or tab that fits to perform specific tasks in a uniform way. While getting started with macros firstly, you have to create a developer tab on your ribbon that will make it easier to add or remove macros to the Excel spreadsheet. Here are some steps that will surely help to how to create or record your macros to the list:

1. Go to the developer tab and choose the macro recording button in the code group. Or you can go with the selection through bottom left corner of your spreadsheet.

2. Now select and write down the name of your macro that should be appearing first. Keep in mind that it should not be with spaces but you can use underscore.

3. After that, choose a shortcut key through which you can apply and select the respective selection without going to the dialog box.

Keep in mind that do not select the random short keys that used to perform other tasks or may be complicated like "Ctrl+Z". Use the combination "Ctrl+Shift+Z" this format will provide you maximum options to design your short key.

4. Next step is where to save your recorded macro; usually people save it on the workbook they are working. But if you want to use this macro for you're the drop down menu.

5. Here you have a description, so write a little about your macro details like what will do with this etc.

6. And at the end by clicking OK

7. After clicking on OK button, recoding of macro just started and while recording, you can adjust some setting before you begin. This includes adjustment of font size, style and color of the cell and column.

8. When the steps are done, stop by clicking on "stop recording" in the code group or the blue

square on Excel window lower left corner.

Quickly master Excel macro development

No doubt macros can make your tasks easier and quick, but its quiet difficult to learn and complex to get started with. If you want to learn and to be master with the use and understanding of macro firstly, you have to gain understanding of the VBA that is a programming language used in office application like in Excel and considered as the basics of macros. To get quickly master with the Excel macro development and use, it is important to learn Excel spread sheet tasks with the basics for VBA.

Preparation to be master with Excel

First of all, you need to get your VBA installation, both versions for window or Mac are available that provide a professional experience of using and learning about the Excel spreadsheet usage. When the installation is complete, you can check it through "ALT+F11" to make sure is that the version is right and VBA window is opened. After that, install your Excel through the control panel and run it to your system. Go to control panel > add/remove software> change Microsoft office > office shared features, and go with the instruction to have your application prepared for working.

Make sure to set a low security level for your macros initially, so it would be able to run the system and have permission to access the system completely. Checking security and allowing it depends on the Excel version you are using on your computer like most used versions are 2007 or 2011.

Check the security for Excel version 2007 first installed and then do to developer > macro security >click the box with "disable all macros with notification." For Mac users firstly they have to check the developer ribbon then set preferences.

Excel for Mac 2011 go to Excel > preferences > ribbon and click on developer, after that set preferences " Excel > preferences > security > Macro security, then click the box "warn before opening a file that contains macros."

VBA editor is almost same for window or for Mac but appears with slight changes like the change in the menu bar from where the main command is actually given.

Prepare for the security threats

It is essential to save your worksheet from the threats of the security and viruses that may harm your system as well as the saved files and documents. While saving macro or codes make sure to not use standard file extension. Like regular files are saved with the extension ".docx entension", always go with the special extension ".docm extension". While you open this version of file it will warn you or enable the macro to run automatically and save your file from the external threats and give you a notification to enable respective macros for the document that may be enabled to keep your document safe.

You can also specify and set the specific or trusted macros location by creating your own trusted location, document or publisher. Just open the security setting in your office application, go with the dialog box > choose trust center then > trust center settings.

Should focus on the code format

While recording or writing macros, the thing which is most important to consider is to keep the focus on the format of writing code. Like

- Sub Macro1()
- Macro1 Macro
- Range("B1").Select
- ActiveCell.FormulaR1C1 = "Hello World"
- Range("B2").Select
- End Sub

The above mentioned format is for all macros recording and writing. If you are recording one macro for a specific column or a row then make sure to stop it first before going to record another macro for other column or row. The signs like an apostrophe, inverted commas, brackets, etc. are a part of the macros and should be well written with an appropriate comment about a code and its use or specifications.

Recording or code generation is a complex algorithm that requires a must understanding of its use and properties that help to perform specific tasks and having particular features that make work even easier.

Make your code editing

Sometimes a little change or editing in an existing code will not only make it easier for you but also helps to address more than one model to perform a task. Like if you have record one code of a particular cell then by editing the column, you can make it for another column or row as well. With the simple copy pasting, you do not need to go through a complete procedure to open developer and create macro again. Simply open a dialogue box and developer tab and select macro one and paste it for other column and save.

Simplifying of daily tasks

We are getting engaged in programming, software,

applications and tools to perform certain tasks in our routine to accomplish our daily tasks. Most of the programs are provide interference and connectivity to make certain things even easier to perform with no time. Like for Excel working or to make your spreadsheet, it is vital to get to know about the codes that make work easy and quick. Excel worksheet is not that much easy for new or a beginner to use and get pro with that to simplify the task within less time. But that is quiet easy with the use or understanding or macros, which provide the specification of shortcuts to apply formulas to add/delete row or column, change font size, color, make highlights, record data, keep worksheet security and much more.

There is a specific set of codes that you have to set just once while starting the working with a particular spreadsheet and with this, you do not need to repeat the task again and again. Just set instruction at once and you can apply it by simply a click or copy pasting. Anyone can learn and know about these codes or macros with instructions and make the task easy to accomplish. Some macros are built-in on the respective version of Excel application that can set on the main tool bar as per the requirement or a person can set customize codes as well with the specific short keys.

Here are some common and useful macros that are commonly used in daily tasks and make things even easier to complete task within less time.

Add serial numbers

You can make your document with a systematic format by using serial number code. Choose the number macros and make your column streamline with the proper numbering, you do not need to apply it for every row or column just select at once and apply to all selective section. Here is the way through which a macro of serial number can be added and apply:

- Sub AddSerialNumbers()
- Dim i As Integer

- On Error GoTo Last

- i = InputBox("Enter Value", "Enter Serial
  Numbers")

- For i = 1 To i

- ActiveCell.Value = i

- ActiveCell.Offset(1, 0).Activate

- Next i

- Last: Exit Sub

- End Sub

Multiple columns

In the routine work on Excel spread sheet, you need to add or delete the columns on your worksheet. For this, go with the use of a certain macro that will help to add multiple columns at once and make your work even easier. Here is the way that will provide a solution, so follow the way mentioned below:

- Sub InsertMultipleColumns()

- Dim i As Integer

- Dim j As Integer

- ActiveCell.EntireColumn.Select

- On Error GoTo Last

- i = InputBox("Enter number of columns to insert", "Insert Columns")

- For j = 1 To i

- Selection.Insert   Shift:=xlToRight, CopyOrigin:=xlFormatFromRightorAbove

- Next j

- Last: Exit Sub

- End Sub

Make your columns auto fits

On worksheet, if you adjust each column, one by one will take time as well as make your work challenging and hectic as well. But the auto fit column macros it will be done quickly, and you do not need to apply one by one manually. Just click on the once and columns and data will arrange automatically. Simply follow the code given below to make your columns auto fits.

- Sub AutoFitColumns()

- Cells.Select

- Cells.EntireColumn.AutoFit
- End Sub

Merge cells

Now you do not to manually merge the cells into one to make your space even more specific. Here is the code that will do it for you just at once. All you need is to select respective cells that you need to merge and simply apply the code and with the blink of an eye, the task is complete. Here is the code:

- Sub mergeCells()
- Selection.Merge
- End Sub

Unmerge cells

As you can merge cells through macros, similarly unmerge code is also available. You do not need to perform task one by one to unmerge hundreds of cells. By applying the code, you can do it just at once. It is one of the most common and popular code that is in use by people who use to work on Excel worksheet and make their work easier for them.

Follow the code in the given format:

- Sub UnmergeCells()
- Selection.UnMerge
- End Sub

Go for document print

Printing document is the most common and famous macros that can make printing more specific to a particular page. Now you do not need to print the whole file or manually do the hectic task if you want to print one page or few out of complete worksheet. Through printing code, you can customize the print range of the page and set your own.

- Sub printSelection()\
- Selection.PrintOutCopies:=1, Collate:=True
- End Sub

Here are some of the macros that make life easier, similarly there are multiple of macros that are commonly used by the people who use to do work on Excel spread sheet on a daily basis and with the help of such short keys and codes perform tasks easily within no time. Some are simple and other may be advanced but with proper instructions, it will easy to follow, apply and to perform tasks.

Work more efficiently with ranges, cells, and formulas

Microsoft Excel macros not only reduce the time but also help to complete task with less effort and make it more accurate for the user. To write your own macros, you do need to be pro with the programming languages and applications as well. Just with the basic knowledge and by following little instructions regarding macros and VBA, you can perform tasks well. To be professional, there are multiple instruction programs available but a person who is in continuous interaction with Excel worksheet and VBA can perform specific tasks easily.

It does basically provide certain range of cells, ranges, columns, and formulas to complete a task quickly. Some codes are by default are part of the Excel program and some can be customized with the respective use and requirement. If you want to create your own macro, even some approaches will help to make it easy for you, like:

Name of macros

While creating your personal macro, make sure to choose a more specific name for it. If you have multiple macros then consider a concise name that can be easily identifiable. As well as a description of the respective macros is necessary that make it useful for everyone who works on that sheet.

The name should not be too long or should be without spaces, you can use symbols and numeric symbols and letters to make it unique, but the maximum characters are 80, so make sure your name should be in the range of 80 characters only.

Start with the home

The start is always with the home cell that is A1 means the very first row and column of your worksheet. No matter where is the cursor when you start recording the macro it will automatically start

with the home and should be with the key
"Ctrl+Home". If you save macro for the worksheet,
then it will limit only with a specific file on which you
are working. But instead of this if you make your own
personalized macros worksheet then you can use this
with other sheets as well.

This will help to organize and streamline multiple
spread sheets by simply combining all worksheets
and generate data report or compile information in
less time. But this will only help if your all spread
sheets contain the same data nature.

Record macros with relative cells

While recording macros make sure not to save the
macros with absolute cell because this will not allow
you to use the code for another cell and only be
specific with a particular selection only. Go always
with the relative cells address because through this,
you can use or make changes in the recorded, and its
ability is not limited.

What's more with VBA programming?

VBA no doubt is a programming language that not only helps to make things easier but also helps to understand and create macros that make your work more efficient and quick with Excel. It is quite easy to get understanding about the worksheet and working on Excel sheet by using codes and a person will be able to get its knowledge to make reports and compile data with the tools, formulas, and codes. Here are some more options that are a part of programming and provide further exploration.

Variables

You can use and alter objects, values, and references for the short-term use in a respective file that change or elements are considered as variables. These variables are used to create a certain code for a limited time period or for the specific tasks.

Here is the code and the way to use it:

- Dim MyStringVariable As String
- MyStringVariable = "Wow!"

- Worksheets(1).Range("A1").Value = MyStringVariable

Branching

The VBA programming is the most powerful tool that provides multiple options and execution facilitation to make your work specific and helps to determine multiple codes to perform specific tasks. Even every code is specific and applies on a certain or specified scenario. You can customize or multiply the task as well as repeat the same operation for multiple times to accomplish the worksheet. By simply copy and paste the code from one box and apply it to others will make it easily duplicate and branching it into further.

Here how to branching the code into multiple one:

- Sub Macro1()
- If Worksheets(1).Range("A1").Value = "Yes!" Then
- Dim i As Integer
- For i = 2 To 10
- Worksheets(1).Range("A" & i).Value = "OK! "

     & i

- Next i
- Else
- MsgBox "Put Yes! in cell A1"
- End If
- End Sub

Final words!

Programming language like VBA is complex to understand but it does not mean that a person needs to be pro with the programming to use it to simply do the daily routine tasks with Excel sheet. In Excel application, this will help to understand and use the macros and to record or customize your macros work sheet. With a quick start and instruction understanding, you can perform tasks easily and accomplish within less time. The task change in font style, size, color, way to highlight, and many more not only make your data compile well as well as helps to generate reports in a systematic way.

CHAPTER 3 – COMMON PROBLEMS AND HOW TO FIX THEM

Excel Macros and VBA seem to be a new technique for most of the professionals out there. It is because of no or less familiarity with the programing, language and macros. It is a matter of fact that whenever we are learning something new, there are possible errors and issues with the recent practices. We sometimes miss out any loops in codes or programing that cause some critical or minor issues with the results.

As a beginner, when you start with the Macros and VBA in Excel, you make so many mistakes but unable to locate them or fix them in real time. To help you with the identification of these common problems or mistakes and their solution here is a set of problems you need to work on. By considering these problems, you can focus on progressive learning and avoid further issues with the overall issues with Macros and

VBA.

Not having developer tools marked

One of the most common issues that every non-programmer Excel user have to face with the VBA macros is the identification of the developer tool. It is normally perceived that VBA and macros are available in the front foot tabs of the Excel. Although, VBA is a built in programing language for Excel and other Microsoft products but still you need to mark the developer tools before getting started with the macros.

Without enabling the developer tool, it is not possible for you to get the options and further navigation for the coding and macro creation. Before getting started further, you need to setup your developer dashboard so you can make additional easy and quick moves with the macros.

Enabling developer tools is not a difficult thing to do, all you need is to go to File > Options > Customize Ribbon. Now move the developer command group in the customize Ribbon from left pane to the right one.

Make sure to enable the check box so you will get the developer tab in the Excel menu and will be able to operate Macros directly from there. Then you can access the code editor window by just clicking the View Code button under controls in the developer menu.

Odd variable tags

When writing the codes or Macros, variables are one of the most important and basic components. You need to define the variables in the first place to get the things done efficiently. In VBA code, writing one of the major issues is the odd variable tags or titles. Most of the beginners are unable to customize the variable tags that cause them to stuck with the odd and complicated variable titles. The automated variable tags are hard to understand, especially if you are new to the coding or working on complex codes. So, it is better to change the titles as per your requirement. It will help you to write the code efficiently.

Remember, the variable titles should be as short as possible and descriptive in nature. When you are coding information, it is necessary for you to have an easy understanding of the codes. It will help you to make some easy moves and move forward with the things. Make sure to keep the variables tag easy to understand by the other programmers as well.

Default sheet titles

Another very basic mistake made by the Excel users is the default sheet titles. Most of the times, sheets are saved as Sheet1, sheet2 and so on. It adds more confusion and keeps the data random and unsorted for the programmers. When you are writing up the macros it is hard for you to evaluate what data is available in which sheet because of no proper tags and titles. In this regard, it is necessary to title the sheets appropriately as per the subject name. It helps in making the real difference in overall coding of the program.

It helps you to refer to a sensible sheet name in your Excel VBA code that will let you connect things properly. It reduces the confusion between the sheets and you do not have to close the macro window to check the sheet name every time.

Poor with loops

For most of the beginners in programing and VBA Excel macros, it is hard to understand the different between looping and breaking when processing a data. When they start writing up a macro, they commonly break the stream instead of creating loops it is not so common and they are not familiar with these options.

In Excel coding and macros, looping is very common because you are commonly processing data with values in the entire column or row so you need to process the data collectively. To handle the huge data, it is necessary to make the use of a loop that will keep things connected properly. Looping is all about providing conditions in coding for actions. There is a use of "if" and "if not" that make coding work for multiple conditions and scenarios.

For the new coders, it is a little difficult to avoid the code brake and keep the loop of the codes to ensure sustainability. But, when you are improving your skills, make sure to have a grip on the looping in Macro coding so you will be able to code things right and will not have to enter multiple codes.

When you are not using the loops, you have to add the codes repeatedly for multiple situations and data conditions. It expands your coding sheet and makes it complicated as well. In this manner, if you need to alter any of the code or there is a mistake in any code then you will have to check a huge list. Looping saves you from the excessive coding and helps you to keep the things summarized.

For the new programmers, it is sometimes difficult to handle the loops and breaks in long codes but to ensure the best skills it is necessary to deal with this situation. By focusing on minor details and working on the basics, you can simply understand the use and proper implementation of the loops and breaks in a code easily.

No use of arrays

Another problem that any beginner programmer faces when they try to process all the variables and data sets under single nested loop is the proper calculation of the data. Dealing with a group of items with similar properties can be tricky and cause you a number of issues. It is difficult to sort such data and get accurate results. Most of the beginners treat such data with respect to variables and it takes so much time to them in sorting out the information.

In fact, if you use the arrays for such similar data then it will be easy to deal with such data and streamline the information correctly. Array is used to deal with different elements of a data with common names. It distinguishes them with different numbers to give you a clear representation of what is available in your array and what not.

Performing the same, calculations on the same number from the same column can lead to the major performance issues and gives you the unidentified error sometimes. Looping through the column and finding the values, every time is a killer process that takes time and a lot of focus as well. If you want to handle the data efficiently, then array is one of the simplest and nice ways to have it.

VBA Macros in Excel are designed to help you with the tasks and calculations. If you are unable to make any progress with the use of these tools and codes then you may not be able to have the best outcomes of the utility. Here using the array function, you will be able to make the smart move with symmetrical data that can drain your energy if you are no using the right tool.

Frequent use of references

No matter if you are using the Visual Basic or VBA, to access some of the features like writing output to a test file of Access database you need to include references. References are a sort of built in libraries contains some resources for you to have support in different acts.

To add the references tab in your developer view all, you need to enable the references and then you can access the references by clicking on tools in the menu. For different versions of Excel, the location of references can be different, so you may have to look for these options carefully.

Here in this dialogue box, you will be able to find out about the references included in the sheet. It is necessary to exclude any unnecessary reference from the sheet as this can waste system resources and cause issues for your further progress. A reduced number of references in your programing project helps you to run the application efficiently and without any prior damage.

Identifying the right action

Another major issue with the beginners in creating macros is the identification of right action. In macro creation, you do have a number of options available in terms of codes. For every possible action, you can have codes with you. All you need is to know the right way of writing the code and move forward with it. To get the right results, you are supposed to take right action.

If you want to work on a single sheet and secure the rest of sheets, then you need to use the macro for securing all sheets other than the active one. If you will secure them, you will have to enter the password every time you will come to the sheet and it will be irritating for you as well. Same as for other tasks you need to get the specialized and specific macro that will make the job easy.

To have the ultimate skill of identification of right action you need to expand your library for the macros and the possible actions you can do. More you will know about the macros and their codes more you can identify the right task. It will help you to make some real advancement in the overall scenario.

Overlapping the macros

For the non-programmer beginners, it is hard to understand the borderline of the macros codes and settings. Initially, many of the beginners mistakenly overlay the macros with each other. It happens when you are using multiple macros without identifying the real purpose or action. If tis about locking the columns then you should use the macro only for the columns, not for the rows. Moreover, there is a macro to lock all columns other than the active one while the other macro locks all the columns of the sheet. You should know what you want to use in the specific sheet.

If you use both macros on one sheet, then you will have to face errors. So, make sure in case if you have entered the double macros or overlap, then go to the developer window and remove the additional or intervening macro from the block. It will help you to restore the normal sheet again and the problem will be resolve.

It is necessary for you to understand the importance of proper and aligned use of macros so you will not have to face issues with the final calculations.

Mixing the codes

When you do not have a basic understanding of the codes, it is not a good idea to get started with it. Using the broken codes or multiple codes to make one file at the beginner level seems to be a waste sometimes. A few programmers can mix numerous codes into one using all the tools in the right place. Not everyone can do the act and have incredible results.

Being a beginner when you try to make something advanced, you may have to face the issues with efficiency and result as well. In this manner, you have to gain all the knowledge about the codes in the beginning so you will be able to make some of the amazing combinations in the end.

Make sure when you are combining multiple codes you will script them right in order to achieve the best results. Sometimes the overlapping of codes can lead to some confusion for the program. In then end all, you will have the errors and issues. It is better to understand the code writing properly and then try to make combinations of codes that will help you to get the ultimate results you need.

The combination codes and macros are helpful if you code them correctly. These complex codes can help you to make some complicated jobs easier and will get the results in one go. So, make sure you are going to have the proper understanding in the first place and then will take the next step to get things right.

Identify the errors

Most of the learning programmers are not aware of the possible errors they can face when coding the macros using VBA. It is one of the confusion that put the things on a standstill in most of the cases. For the beginners, it is necessary to keep track of possible and common errors and their identification along with learning about the codes and macros. The identification of the error helps them to evaluate its solutions.

When you know that what can cause you a specific

error you will have the way to resolve it using the possible way out. Make sure you are going to make the use of your skills in keeping both ends matched. At one side, if you are learning about the codes then, on the other hand, you should have the idea about the errors and their reasons. The reasoning helps you to deal with the error efficiently.

In case of an unknown error, you can take the built in help from the VBA developer center in the Excel. It also helps you for the identification of the errors and you can lead to the results for the solution.

Do not use multiple files

If you are going for the Excel macros using VBA, then you need to specify the file you are working on. Sometimes opening multiple Excel books at the same time can lead you to the confusion of the coding. It can be a fault on your side as well. So, make sure you are using only one file at a time.

In case you need to have the multiple files, open as you are taking figures from the other sources then keep them personalized. By locking the file and its worksheet, you will not be able to make any random or blind changes to the other file. It will help you to apply the macros only on the selected file in a specific time zone.

Specify the worksheet

Working on multiple worksheets can risk you sometimes with macro coding. If you are unable to select the worksheet properly, then there is a chance that macro will apply to a wrong sheet. In this regard, it is necessary to specify the sheet correctly.

To select the sheet, you can simply lock the rest of sheets to specify the action sheet. You can even use the macro code to lock them by keeping the active one unlocked. It can help you with the proper application of the macros on the sheets and get the desired results in the end. Make sure you are going to make the right selection of codes so you will get the best results.

Uncheck the auto application of macro

When you are working on Macro VBA frequently and

using some of the handy macros to help you with the routine work then you have the option to make these macros applicable automatically. Selecting the auto option will help you to use these macros in multiple files. You are not supposed to enter these codes repeatedly. It is one of the best helps that you can have with VBA coding.

On the other hand, this can cause you issues and errors. There can be files in which you do not want to use these macros and wants to keep the things normal. Then you need to uncheck the auto application of macro in the file. It will help you to avoid any issues with the new file. In fact, you can adjust some of the specific conditions that will lead to the application of a specific macro. To apply all these tricks, you need to have command over the macros and their application in Excel.

Bugs and debugging

When starting up with the macros and VBA, you need to understand the type of common bugs that can attack your system or software. One of the common errors is the syntax error that violates the grammatical rules of the language. It is cased when there is a minor fault or use or improper value in the coding ad it affects the overall execution.

The best thing about the error is its automatic identification. When you enter the code, it will blink and flash on the screen when there is a syntax error in there. For instance, the string values should be enclosed with double quotation marks in VBA, if you use the single quotation marks then there will be an error showing you the red warning. You need to select OK and change the single quote sign with the double quote and it will be sorted.

Most of the time, writing up the codes can be tricky with this and hectic as well. You might do not know where you have made a critical mistake and how it needs to be treated.

The other common errors are the runtime errors that are difficult to evaluate and find out as well.

Apparently, everything seems fine but when it comes to the application of code it does not execute the desired results. These issues are frustrating because you never know the reason. In this error, there is not syntax error and you have all the codes as per the guidelines and language mechanics.

To find out about this bug, you need to open the Visual Basic Editor and check the error in the code that is running and failed. You will find out a little change in the Value; it will be showing as ValueX instead of Value. You need to change it to Value and then click the little green play button under the Debug menu. With these few steps, it will be sorted and you will get the code running smoothly.

CHAPTER 4 – TOP TIPS FOR BEGINNERS

VBA seems to be easy to understand and practice programming language out of the other available options. For the programmers, it is an easy task to learn, understand and practice the VBA language for the Excel macros coding. But, for the non-programmer, it is important to learn it individually to work on the macros and have some of the advanced options in programming and coding.

We cannot ignore the importance of macros in Excel and their role in increasing the working efficiency of the professionals at the same time. The ease of understanding makes it easy for people to learn the language and understanding coding. But, sometimes even for the beginners, things get a little difficult and out of control. They may not find it helpful to understand and practice the codes and language right.

Other than the approach or ability to understand and deal with the language or code, some people od make mistakes. Beginners have some of the flaws in their initial practices that cause them failure or issues related to the learning and practicing as well.

To make your learning experience of VBA and macros easy and accessible, here are some top tips that help you to make some real quick and efficient moves with the training. Other than knowing the importance of programing in macro coding for Excel, you need to know some additional information that will help you in a longer run.

Evaluate your potential

Before getting started with the VBA and Excel macros, you need to know your potential. It is not necessary that everyone can understand the codes and programming language. A few people are not comfortable with codes, so they need another way out for the problem. This initial evaluation helps you to understand whether you are going to learn the coding or making your own reserves for good.

You have two options to work with VBA and macros, one is to learn the programing and VBA language that will help you to create macros. It is all about getting the best skills for your personal improvement and achieving the ultimate results. The second option is to gain basic knowledge about VBA and macros so you will be able to real them and evaluate their purposes. However, you are not going to write the codes, in fact, you will maintain a gallery of your own containing multiple codes.

The macro coding and examples are available out

there, all you need is to collect them with you for the future use and it will help you. Here one thing is compulsory and that is the basic knowledge of the coding and reading macros.

Boost up your programing knowledge

If you want to achieve the status of an advanced and professional coder with Excel macros, then you need to have a real boost in programming knowledge. You should know the problems you may have in learning the codes and language and then sort the problem to achieve the best results. It seems a little tough but not impossible. All you need is to reach the right resources.

There are online sources available as well as help center of the Microsoft Excel helps you to know more about VBA and get the answers to the possible issues you are facing in your overall progress.

Learn the basic format

The best way to learn nay programming language or coding is to understand the basic format. Every language has its own rules and formulas. When you have the idea about these basics it is simple to move forward with it. If you are planning to copy some examples and get started with these codes in creating macros and moving forward, then it will not pay you off that much.

Copying the available macro-codes will be just temporary support that you can have; until you are not able to understand or create one for yourself, it's not worthy. Make sure you will learn the basics and go through all the common things are necessary to look into. It will simply help you to make some of the amazing progress with the coding. When you know the pattern, you can fill up the codes and words in there easily.

Start with examples

Learning always starts from the demonstration and tutorials and in this advance-programming era you can find out numerous examples and visual demonstration that helps to learn anything. Like if you are going to start working or learning VBA programming and want to create macros or to learn about its usages, it is preferred to go with the examples. Like small tutorials, video clips are available that will not only provide knowledge about the respective code as well as you can find out how to apply and use it to perform certain tasks. In examples, you can also find out the code writing patterns and ways to write it in a correct form. You can pick up the relevant one and should apply it to your worksheet. You can select the code from the default setting of the application as well as make your own customize spreadsheet that will remain save and you can apply the save codes on the multiple worksheets as well at once.

Solve the simple problems

Using Excel is complex and quiet difficult as compared to any other application because here every task needs a specific code or formula that can reduce the working time. Other than the use of macros or VBA, the use of Excel is quiet time consuming and hectic and you have to perform every task manually. The best way is to go with the simple and easy tasks first then gradually go towards the complex one. In simple tasks like how to change the color of text, how to add columns, alter rows, merge or unmerge the cells, highlight the column or rows and many more. Firstly, learn about the small tasks and simply apply them into your worksheet and make a practice of it. Through practice, you can get the understanding and will help to remind the short keys and there use as well.

Do not jump for complex issues

VBA Excel macros are not just about handling some of the basic issues and problems. The technique helps in sorting data within no time and reducing your formula efforts to achieve accurate calculations and results. The codes are ideal to be used with the complex tables, larger data and for bigger purpose as well.

To reach the complex level of coding, it is necessary

to learn the basic one at first. Most of the beginners make one mistake and that is of jumping to the complex codes and matters. It confuses them more and makes things harder to achieve. As they are unable to understand the basic structure of codes, so it is hard for them to achieve the desired solution from these complex problems.

Stay focused to solve errors and issues

While doing the macro coding, you can have several errors and bugs just like any other programing language. These errors are sometime system generated and sometimes happen due to coder negligence. To deal with these errors and bugs you need to be focused on job. Make sure you are going to review all the codes every time properly.

A proper review helps you to avoid the chances of issues with the coding. Moreover, keep yourself updated from any progress and update in the language, software or macros as well. Sometimes, a certain bug in the server can cause you to face issues with the coding. If you have all the updates, you can avoid any harm.

Develop data understanding approach

It is not necessary that every tool or macro can be used for any data. Other than the sheet, oriented macros there are certain macros used for data processing and evaluation. You need to understand the nature of coding and data at the same time. It will help you to evaluate the right data using the right code.

If you are using the single column macro for combined data or applying it wrongly to the data then you will not get the right results. It can result you to have any kind of error or issue with the calculations as well. Make sure you are going to have a proper understating about the data and then decide to apply the macro coding on it accordingly.

CHAPTER 5 – SOME USEFUL EXAMPLES FOR BEGINNERS

It is a fact that when we have some of the easy examples available, we can do things better. If you are learning the VBA Macros for Excel, you do not need to hurry. It will take a little time for you to understand the codes, statements and other technical points. All you need to do is to focus on the small tips and examples. Here we present you some of the useful, basic and easy examples that will lead you to practice and being pro at it.

It is always good to start with some of the basic examples so you can understand how things are working in an overall scenario. Once you are done with it, you need to move forward with the further advancements. The basic practices help you to make some additional improvements.

Practice the examples properly

When you want to make some real moves with the VBA Macros in Excel, you need to practice the basic codes properly. These examples help you to understand the basics and then move forward with the advanced codes and commands for actions. Make sure you are going to practice these examples properly before going further with the advanced codes and options.

Incorporate your combinations

Once you have the idea about how to write codes and will observe the basic examples, then you can move forward for the personalized combinations. These combinations will help you to advance in so many things. You can prepare some of you won personalized codes that will help you to make work easy and quick.

Look for the possibilities

To get some of the personalized codes, you need to evaluate the possibilities in Excel macros. When you will have the idea about how things work so you then it will be easy to use the codes and macros as per your needs. The Macros in VBA offers you the liberty to expand their coding and usage as per requirement. You have question then you can find solutions and answers properly. It takes a bit of practice and then you can make things fine for you.

Some useful examples for VBA Beginners

Here we bring you some of the easy and basic examples that help to understand the Macros and extend the possibilities for usage as well. Have a look at these resources to make some extensive moves for the ultimate results.

1. Make All Worksheets visible at One Go

Working on multiple sheets is hectic but by hiding them for goo helps you to make things symmetrical and easy as well. But, eventually, you need to unhide these sheets so you can look up to a compelte data. Now, to unhide all the sheets one by one will take a lot of time and energy, so its good to make the use of easy macros and get the job done quickly.

You need to type:

- Sub UnhideAllWoksheets()
- Dim ws As Worksheet
- For Each ws In ActiveWorkbook.Worksheets
- ws.Visible = xlSheetVisible
- Next ws
- End Sub

The code will simply change the visible property of the worksheet to visible and let you to have all the sheets in the task bar at once. You can save a lot of your time at one go.

2. Keep the active sheet visible

When working on a report, it is hard to focus on work when you have so many sheets on the dashboard. It can confuse you and sometimes will cause you to have issues with the readings and numbers as well.

To deal with this situation you can use an easy option to write the macros and hides the rest of worksheets other than the active one. It will help you to keep the active sheet on board and hide the rest of the inactive sheets easily.

The code for the macros is:

- Sub HideAllExceptActiveSheet()
- Dim ws As Worksheet
- For Each ws In ThisWorkbook.Worksheets
- If ws.Name <> ActiveSheet.Name Then ws.Visible = xlSheetHidden
- Next ws
- End Sub

The macros code will help you to get start with the report and can give you some assistance in the overall project as well.

3. Arrange the worksheets alphabetically

VBA is not just for the advanced actions and

commands, in fact, you can make a real difference in the basic tasks using the macros. This code helps you to sort out the worksheets alphabetically in no time. The code helps most of the times in reports when you have to manage things with names and in alphabetical order. You can align the sheets as per employee names or products names using the easy code.

The code for the macro is:

- Sub SortSheetsTabName()
- Application.ScreenUpdating = False
- Dim ShCount As Integer, i As Integer, j As Integer
- ShCount = Sheets.Count
- For i = 1 To ShCount - 1
- For j = i + 1 To ShCount
- If Sheets(j).Name < Sheets(i).Name Then
- Sheets(j).Move before:=Sheets(i)
- End If
- Next j

- Next i
- Application.ScreenUpdating = True
- End Sub

4. Quick protection for all worksheets

When we work on some massive worksheet with many codes, numbers and figures that are complex and confidential, then you should protect these sheets. It will help you to avoid any unauthorized access or edit in your worksheets and you can rely on the compilation as well. The macros let you to have a specific password in the code and you can access the sheet using this password.

Code for this macros is:

- Sub ProtectAllSheets()
- Dim ws As Worksheet
- Dim password As String
- password = "Test123" 'replace Test123 with the password you want

- For Each ws In Worksheets

- ws.Protect password:=password

- Next ws

- End Sub

Remember to mark the password carefully so you will end up having the secured password and will not have to face any issue further.

5. Unprotect worksheets at once

Accessing the protected worksheets is sometimes difficult when you have some or all the worksheets protected with a password. If you are switching between the multiple sheets, it is not comfortable to switch the sheets and enter passwords every time. By using this code, you can unprotect the worksheets and get the work done.

For the macros you need to type this code:

- Sub ProtectAllSheets()

- Dim ws As Worksheet

- Dim password As String

- password = "Test123" 'replace Test123 with the password you want

- For Each ws In Worksheets

- ws.Unprotect password:=password

- Next ws

- End Sub

Remember you need to enter the same password as it was entered to protect the sheets. In case you do not add the same password, it will give you errors in macros and code will not work on the sheets.

6. Make all rows and columns visible

Sometimes when you are using some external source files to work, you can find it hard to locate some of the rows or column in the worksheet. It is because the columns or rows are hidden in the sheet and you are unable to see these. Here is a macros code to make all the hidden rows and columns visible in your worksheet.

You need to write the following code:

- Sub UnhideRowsColumns()
- Columns.EntireColumn.Hidden = False
- Rows.EntireRow.Hidden = False
- End Sub

It can be a great help when you are dealing with some complex files or sometimes mistakenly, you hide the columns or rows in your worksheet.

7. Quick unmerge of all cells

When working on your worksheet, you need some more space that will help you to enter data and keep the record of stuff. It is helpful when you are making just simple records. On the other hand, it does not help when you want to sort the data. Merged cells are complex to handle with formulas. While working on a sheet it seems difficult to unmerge the cells but using the following macros, you can do it easily:

- Sub UnmergeAllCells()
- ActiveSheet.Cells.UnMerge
- End Sub

The command will help you to get the cells unmerged all at once within a few clicks.

8. Saving worksheet with a time stamp

For longer projects, it is necessary to keep track of timely progress. Most of the times, it takes a lot of your time to keep track of progress whenever you save a file. Macros make it easy for you to save the worksheet with a time stamp in its name automatically. All you need is to save it every time after using the following code:

- Sub SaveWorkbookWithTimeStamp()
- Dim timestamp As String
- timestamp = Format(Date, "dd-mm-yyyy") & "_" & Format(Time, "hh-ss")
- ThisWorkbook.SaveAs

"C:UsersUsernameDesktopWorkbookName"

& timestamp

- End Sub

The time specification helps you to determine the progress and also keep track of data used in the file recently. Moreover, you can save a lot of energy and know how much closer you are to the ultimate completion of the project. One thing you need to consider is the specification of folder location and the file name when it enters the code. You need to consider the destination folder of your choice for the file to get it saved in the right place.

9. Saving each worksheet as PDF separately

Excel offers you to work on multiple sheets at a time in single file. You do not have to open a new file, again and again, to get the work done. It saves a lot of space, time and energy as well. Sometimes, with your final project when you are working on different months, years, or deaptrtments and need to save them in a saperate PDFs, then it's a problem.

It can take you hours and a lot of efforts to find a way and save all the sheets in a saperate PDF file. While using the simple code you can actually make it up and can save the sheets into PDFs easily.

- Sub SaveWorkshetAsPDF()
- Dim ws As Worksheet
- For Each ws In Worksheets
- ws.ExportAsFixedFormat xlTypePDF, "C:UsersSumitDesktopTest" & ws.Name & ".pdf"
- Next ws
- End Sub

The code is workable for worksheets only not the chart sheets. Moreover, you can change the destination folder in the code to save files as per your need. Remember, all the files will have the sheet name as the title, so you do not need to sort these files independently.

10. Converting formulas to values

Using a number of formulas on a worksheet is common for professionals and it is the part of their daily routine. But, converting these formulas into values seems to be a difficult task if you have to do it one by one. It takes time, energy and focuses on having the precision in the job. Using the VBA macros, you are able to convert the formulas into values within no time.

- Sub ConvertToValues()
- With ActiveSheet.UsedRange
- .Value = .Value
- End With
- End Sub

11. Lock cells with formula

Commonly when working on sheets, we have to face issues with the cells containing formulas. One mistake and you may temper the cell with formula and your whole sheet turns into nothing. It is helpful to lock the cell with a formula so you can work properly. It helps you to save time and possible risk of damaging your calculations.

- Sub LockCellsWithFormulas()
- With ActiveSheet
- .Unprotect
- .Cells.Locked = False
- .Cells.SpecialCells(xlCellTypeFormulas).Locked = True
- .Protect AllowDeletingRows:=True
- End With
- End Sub

12. How to Sort Data by Single Column

Data sorting is an important feature and Excel helps to make a sorted data for you. You can arrange it in a specified column and row that will help to make your information uniformed. Here is the code that will help you to sort data easily by using a single column.

- Sub SortDataHeader()

- Range("DataRange").Sort Key1:=Range("A1"), Order1:=xlAscending, Header:=xlYes

- End Sub

In the code, some features and parameters were used that should be noticed while using the code for data sorting like name range. As well as the column in which you want to sort the data and you have to mention that whether you want to sort data in ascending or descending order or the data header as well. Like in above example "key1" shows the data set in which data will be sorts, "order" represent the order like ascending or descending order and in "header" you have to mention about header availability of your data.

13. Sort data by using multiple columns

Instead of single column data sorting, you can also use VBA in Excel worksheet to sort data in multiple columns as well. This will help to keep the record separate with respect to each header like to records sales of different stores that have different locations. If you have a particular dataset, then apply VBA by using the following code to sort it in multiple columns.

- Sub SortMultipleColumns()
- With ActiveSheet.Sort
- .SortFields.Add   Key:=Range("A1"), Order:=xlAscending
- .SortFields.Add   Key:=Range("B1"), Order:=xlAscending
- .SetRangeRange("A1:C13")
- .Header = xlYes
- .Apply
- End With
- End Sub

14. Get the text part only from Excel worksheet

As you can get the numeric part from the worksheet by using VBA, you also can draw and get the only text part from your worksheet as well. By using macrons set the text part only instead of numeric part and get the required column. Following code is use and customized to get the text part out of your Excel worksheet.

- Function GetText(CellRef As String)
- Dim StringLengthAs Integer
- StringLength = Len(CellRef)
- For i = 1 ToStringLength
- If Not (IsNumeric(Mid(CellRef, i, 1))) Then Result = Result & Mid(CellRef, i, 1)
- Next i
- GetText = Result
- End Function

15. Get only Numeric Part from Excel

You can customize your dataset and get the required column only, like if you want to extract only the numeric part from the worksheet then create the

=GetNumeric, function by using your VBA in your worksheet. You can create and add this code on your personalized worksheet that will help to choose it easily for the next time as well.

- Function GetNumeric(CellRef As String)
- Dim StringLengthAs Integer
- StringLength = Len(CellRef)
- For i = 1 ToStringLength
- If IsNumeric(Mid(CellRef, i, 1)) Then Result = Result & Mid(CellRef, i, 1)
- Next i
- GetNumeric = Result
- End Function

16. Highlighting Blank Cells using VBA

While doing the work on the worksheet you often use to highlight the blanks cells, so it's better to use macrons instead of going to the dialog box to search for the respective. Customized your quick search toolbar with this macro to quickly access and highlight the blank cells. Here is the code through which you can create and save it on your personal workbook.

- Sub HighlightBlankCells()
- Dim Dataset as Range
- Set Dataset = Selection
- Dataset.SpecialCells(xlCellTypeBlanks).Interior .Color = vbRed
- End Sub

In the above code, specifically red color is in selection to highlight the blank cell, but you can customize and select any other color of your choice to make your blank cell highlighted.

17. Cells With Comments

On working on your worksheet, you can use the highlights as well as mentioned particular comments on the respective or selective row as well as on column. This will make to remind about something important to you can make notes to specific changes with a report as well. Here is the code that uses to highlight all cells and collectively mentioned comments on the highlighted cells.

- Sub Highlight Cells With Comments()
- ActiveSheet.Cells.SpecialCells(xlCellTypeCom ments).Interior.Color = vbBlue
- End Sub

The mentioned code is with respect to the blue color you can change the color of your choice to highlight by changing the color in the code.

18. Change the Letter Case into Upper Case

While doing working on the Excel sheet, you can make your worksheet more effective and precise as per the requirement and working as well. You can easily change the text format like the letter case of the respective and selected cells into the upper case. Following code will help to perform this task quickly. At once you can use Ucase for upper case text and

LCase for lower case letter. Here is the code for the change letter to upper case.

- Sub ChangeCase()
- Dim RngAs Range
- For Each RngInSelection.Cells
- If Rng.HasFormula = False Then
- Rng.Value = UCase(Rng.Value)
- End If
- Next Rng
- End Sub

19. Refresh Pivot Tables

If you are working on a workbook, then it is necessary for you to keep check and record on the simple and common use code. They will not only help to make your work precise as well as easy for you to complete your tasks. If you are working on a workbook that has more than one pivot table, then by applying this code, you can refresh them all just at once. Here is the code as well as a way to use the code for the respective tasks.

- Sub RefreshAllPivotTables()
- Dim PT As PivotTable
- For Each PT InActiveSheet.PivotTables
- PT.RefreshTable
- Next PT
- End Sub

20. Misspelled Words highlighted

When you are doing work on Microsoft word or PowerPoint, they have their specified tool to check misspelled words by highlighting it. Excel does not have a specified tool to check spellings automatically. No need to worry about you can use code to run spelling checking by pressing F7 key other than that no highlight view that mentioned the spelling mistakes. You can apply and use the following code to highlight the spelling mistake in the respective column or row by simply following the way. This will highlight the names and terminology that may be specified for a particular product or event as a spelling error, so you have to keep eye on it while checking.

- Sub HighlightMisspelledCells()
- Dim cl As Range
- For Each cl InActiveSheet.UsedRange
- If   Not Application.CheckSpelling(word:=cl.Text) Then
- cl.Interior.Color = vbRed
- End If
- Next cl
- End Sub

21. Worksheets protection in the Workbook

If you are doing working on more than one worksheet and looking for the solution to protect the data or sheet, here are codes that will not only assist you in that manner but also make work easy for you. You can limit unauthorized access to your personal and professional data, codes and information. These macrons make your document secure and protected with a password or special characters and only you can have access to that. Here are some codes mentioned below those are useful and effective to use and protect your worksheet in a workbook with just one go.

- Sub Protect All Sheets ()
- Dim ws As Worksheet
- For Each wsIn Worksheets
- ws.Protect
- Next ws
- End Sub

The code will go separately for each of your worksheets and apply the procedure of protection to each of file one by one. For the case, if you want to apply security to all sheets then go with the use of ws.Unprotect instead of ws.Protect in the code.

22. Insert a new Row in the selection after Every Other Row

Through this, working on a worksheet is quite easy and provides all available quick solutions to every question. Like while working, you want to add a new row in the required selection after every other row and just go with this code. Knowing this quick code will help to make your worksheet precise, effective and you will be able to data in the sheet if you missed or you want to add for further. Just select the particular range where you want to add a row and apply the following code.

- Sub InsertAlternateRows()
- Dim rng As Range
- Dim CountRowAs Integer
- Dim i As Integer
- Set rng = Selection
- CountRow = rng.EntireRow.Count
- For i = 1 ToCountRow
- ActiveCell.EntireRow.Insert
- ActiveCell.Offset(2, 0).Select
- Next i
- End Sub

Same code will modify with respect to add column instead of row in the selection area on your worksheet.

23. Insert Date & Timestamp in the Adjacent Cell

A timestamp helps to track activities and people use it keep the check on the updates. It's like if you want to know about the time when a particular file was created when the data was entered or record the time of particular expense as well as wish to see the date when the file was recently updated. While making a worksheet insert this code in a adjust cell and keep a record on a particular date and time about the edited entry. Here is the code and the way to apply it on your work sheet.

- Private Sub Worksheet_Change(ByVal Target As Range)
- On Error GoTo Handler
- If Target.Column = 1 AndTarget.Value<> "" Then
- Application.EnableEvents = False
- Target.Offset(0, 1) = Format(Now(), "dd-mm-yyyyhh:mm:ss")
- Application.EnableEvents = True

- End If

- Handler:

- End Sub

To apply the code on your worksheet inserts it into the worksheet window, not in the module window. For applying double click on the VB editor on which you want to apply then copy and paste the code on the respective window sheet. The code has a line targeted, so keep in mind that the when you entered the data code will work accordingly.

24. Highlight Alternate Rows Selection

Highlighting will always help to prominent the content that seems to be essential or a person wants to keep prominent. It will not only looks good but also increase the readability of the file or information. On the file, it will help to highlight the information that a person wants as well as in document print it will draw attention to the specific selection. You can use the code to highlight a particular row or column that is important to you. Here is the code and the way to use it to highlight particular selection.

- Sub HighlightAlternateRows()
- Dim MyrangeAs Range
- Dim MyrowAs Range
- Set Myrange = Selection
- For Each MyrowInMyrange.Rows
- If Myrow.Row Mod 2 = 1 Then
- Myrow.Interior.Color = vbCyan
- End If
- Next Myrow
- End Sub

You can simplify it by using the particular code of color for highlighting as well, like vbCyan in the code. You can use other colors as well in a way like vbRed, vbGreen, vbBlue respectively.

VBA is the powerful and helpful tools that will not only make your work easy and convenient as well as you can get quick access to data and information. Yu can sort out your problems and complete your tasks with no time. Working on Excel is not that much easy but with the help of codes and macros you can easily crash the points that will make it easy for you to arrange, sort, gathered and streamline information or data and generate quick outcomes out of it. So, learn and apply the macron and make your work quick and efficient with the Excel worksheet like never before.

CONCLUSION

Microsoft Excel is one of the commonly used and easy to access tools for beginners and professionals to do their daily tasks. Other than keeping records, it helps to run formulas, evaluating and sorting data and making reports. No one can deny its importance and efficiency when it comes to performance. An add-on of VBA of Excel and Macros makes it even better option for the beginners and professionals as well. It makes things streamed in the best form.

With the help of an additional language such as VBA, it is helpful to perform some complicated and bundle tasks with the help of Excel. The language is easy to understand and macros can be practiced commonly to increase the work efficiency. The built-in library helps to keep the record of frequently used macros that helps to sort the data and perform multiple tasks efficiently.

To get the best out of VBA Excel Macros, you need to learn the best of its skills that help you to be a competent professional in the field. Learning it for the beginners and non-programmers is not a difficult task at all. All you need is a little resource, practice and a good understanding of codes. Moreover, you should know the proper use and placement of these codes in the right place. Eventually, you can get some of the macros as resources and examples that help you to increase practice and develop more macros of the same nature.

Made in the USA
Middletown, DE
17 February 2020